The Library of Cells™

Plant and Animal Cells

Understanding the Differences Between Plant and Animal Cells

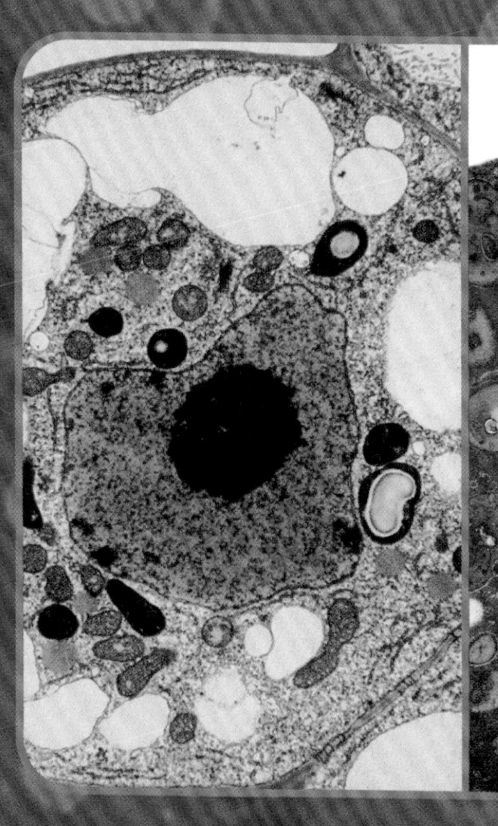

Judy Yablonski

The Rosen Publishing Group, Inc., New York

To Michael Yablonski, who taught me everything he could about science

Published in 2005 by The Rosen Publishing Group, Inc.
29 East 21st Street, New York, NY 10010

First Edition

Library of Congress Cataloging-in-Publication Data

Yablonski, Judy.
Plant and animal cells: understanding the differences between plant and animal cells/by Judy Yablonski.—1st ed.
p. cm.—(The library of cells)
Includes bibliographical references (p.) x.
ISBN 1-4042-0324-9 (library binding)
1. Plant cells and tissues. 2. Animal cell biotechnology. 3. Cell differentiation.
I. Title. II. Series.
QK725.Y33 2004
571.6—dc22

2004016655

Manufactured in the United States of America

On the cover: This transmission electron micrograph (TEM, *left*) shows the parts of a typical plant cell: a nucleus, chloroplasts, mitochondria, cytoplasm, vacuoles, and cell wall. The colored TEM *(right)* shows a microscopic image of *Naegleria fowleri*, an amoeba found in freshwater, under-chlorinated swimming pools, and soil. Although human infections from the protozoan are rare, they are fatal in most cases.

Contents

	Introduction	4
Chapter One	The Building Blocks of Life	6
Chapter Two	Plant Cells	11
Chapter Three	Animal Cells	18
Chapter Four	Plant Cell Functions	28
Chapter Five	Photosynthesis and Cellular Respiration	34
	Glossary	42
	For More Information	45
	For Further Reading	46
	Bibliography	46
	Index	47

Introduction

Imagine that you are a witness in a police lineup, but you are required to do your identification through a microscope. Now, imagine that the suspects are as follows: a porcupine, a potato plant, a squirrel, and an avocado. It's your job to pick out the plants from the animals. After learning about the differences between plant and animal cells, you will be able to identify them from their unique characteristics.

All living things, whether they are plants or animals, are made up of cells. Cells are the smallest, most basic units of life. Cells are the building blocks that make up every inch of your entire body. As it takes many bricks to build a wall, it takes trillions of cells to make the tissues, muscles, skin, organs, and blood that is a part of every living person. However, you can only see cells under a microscope.

Some organisms have only one cell. For example, germs, also known as bacteria, are unicellular. In fact, single-celled organisms are all around us, but we can't see them with the naked eye. Although single-cell germs are likely on your body as you read this, have no fear. As long as you wash your hands before eating, even if they don't look dirty, it is unlikely that the germs on them will enter your body.

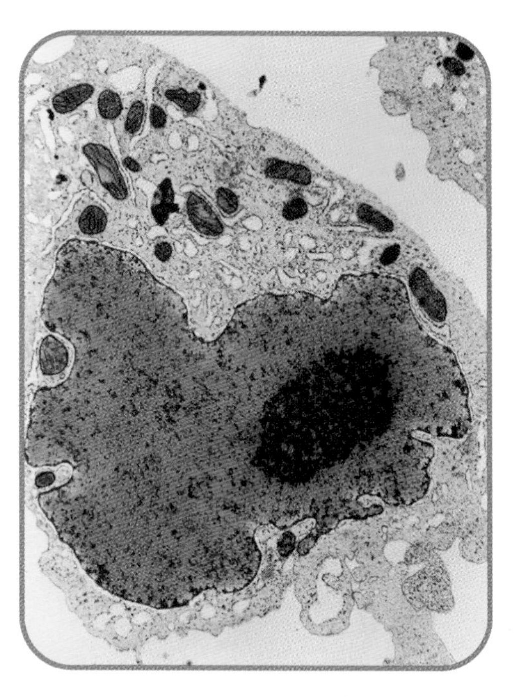

This colored transmission electron micrograph (TEM) shows a typical animal cell. The cell's nucleus, colored pink, is where genes are stored in the form of chromosomes. Inside the nucleus is its active center, the nucleolus, which is colored brown. The cell's cytoplasm is green, and the free-floating brown organelles are its mitochondria, the centers of cellular respiration.

Large living things, such as humans or trees, are multicellular. Such multicellular organisms can be made up of millions, if not trillions, of cells. Cells come in all different shapes and sizes. They can be long and thin, small and round, and they can even be shaped like kites. The different shapes and sizes allow cells to perform their specialized functions. Your muscles are made of cells that help you move, and your blood is made of cells that carry oxygen around your body. Plants have cells that force water up their stems and others that protect the outside of the plant, just as your skin cells provide an outer coating of protection for your body.

Chapter One

The Building Blocks of Life

Despite the differences between cells that perform different functions, plant and animal cells have similar components. All cells, whether plant or animal, have structures that can be found inside every cell. These structures are even tinier than the microscopic cell itself.

All eukaryotic cells have a nucleus, which is like the cell's command center. Eukaryotic cells (cells that contain a nucleus) also contain organelles ("little organs") surrounded by a cellular membrane. This membrane is like an outer "skin" that filters material in and out of the organelles. The cells of protozoa, algae, fungi, plants, and animals are all eukaryotic, while prokaryotic cells, those cells without a true nucleus, such as bacteria, have no compartmentalized organelles.

The Nucleus

The nucleus is the control center of the cell. If a cell were an airplane, the nucleus would be the cockpit where the pilot sits. It is located in the center of the cell, and it stores genetic information known as deoxyribonucleic acid, or DNA.

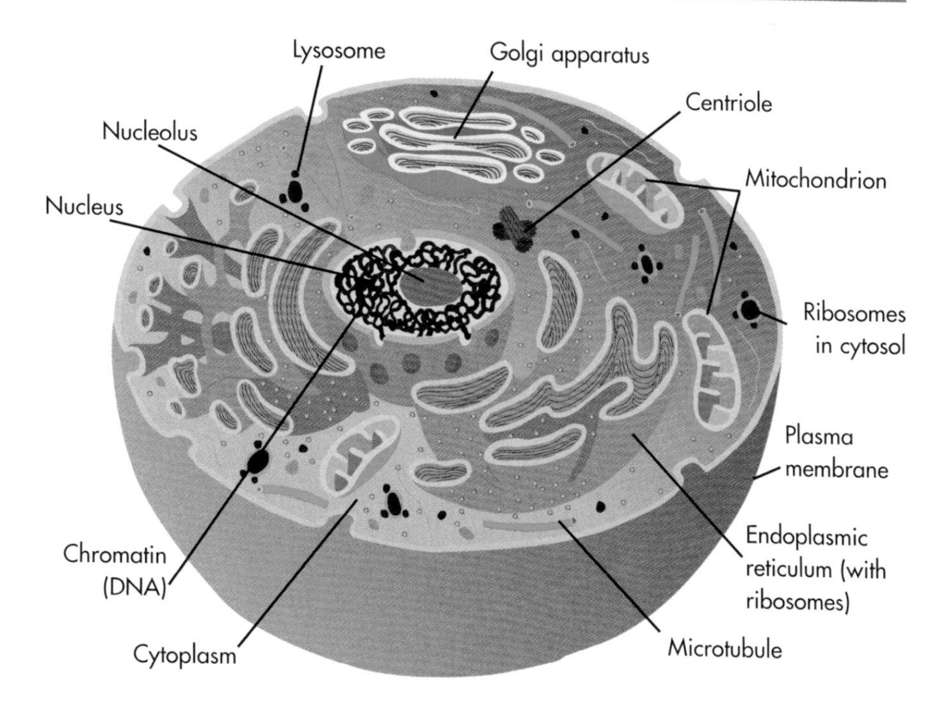

This is a diagram of the parts of a eukaryotic cell. All animal cells such as this one contain a nucleus, where genetic information is housed; the interior of the nucleus, the nucleolus; and the cell's other organelles surrounded by a jellylike cytoplasm. Other organelles pictured are the cell's mitochondria, endoplasmic reticulum, Golgi apparatus, lysosomes, ribosomes, microtubules, and centrioles. All eukaryotic cells are protected by a semipermeable plasma membrane.

The nucleus is wrapped in a membrane with pores or holes, which lets the genetic information pass between it and the rest of the cell. It is the largest structure in most cells, and is often the part of the plant or animal cell that is most clearly visible when seen under a microscope.

Inside the nucleus is a round body called the nucleolus. The nucleolus produces messengers, which are transported out of the nucleus to other parts of the cell. The ribosomes receive and synthesize proteins according to the instructions sent by the nucleoli (plural of nucleolus). The ribosomes are tiny, round particles that often attach to the long stringy structure called the endoplasmic reticulum, or ER. After the proteins are processed by the ER, they are then either used by the cell or exported through its outer plasma membrane.

Other Parts of Cells

Cells also have a built-in security system called the Golgi apparatus. The cells' Golgi apparatus is like a security system that patrols the cells. The Golgi apparatus wraps around substances and either stores them for later use or forces them out. As part of the rejection process, a part of the Golgi apparatus that has the substance wrapped in it will break off. This sac, and its rejected inner substance, is then transported to the plasma membrane.

Most of the cell is made up of cytoplasm. Organelles float freely in this jellylike substance. The organelles each have their own specific role in keeping the cell alive. Some organelles make products for other cells to use, while others deal with wastes. Each organelle is controlled by information generated from the nucleus.

For example, the large organelles that control respiration for the cell are called mitochondria. Once the proteins in the outer cell membrane have chosen the sugars (glucose) they will let into the cell, mitochondria help process those sugars.

During cellular respiration, glucose and oxygen react in the mitochondria to produce the energy needed for the cell to carry out its functions. In this way, the mitochondria are like the "powerhouses" of the cell. Mitochondria are usually found in abundance in cells that use a lot of energy, like liver and muscle cells.

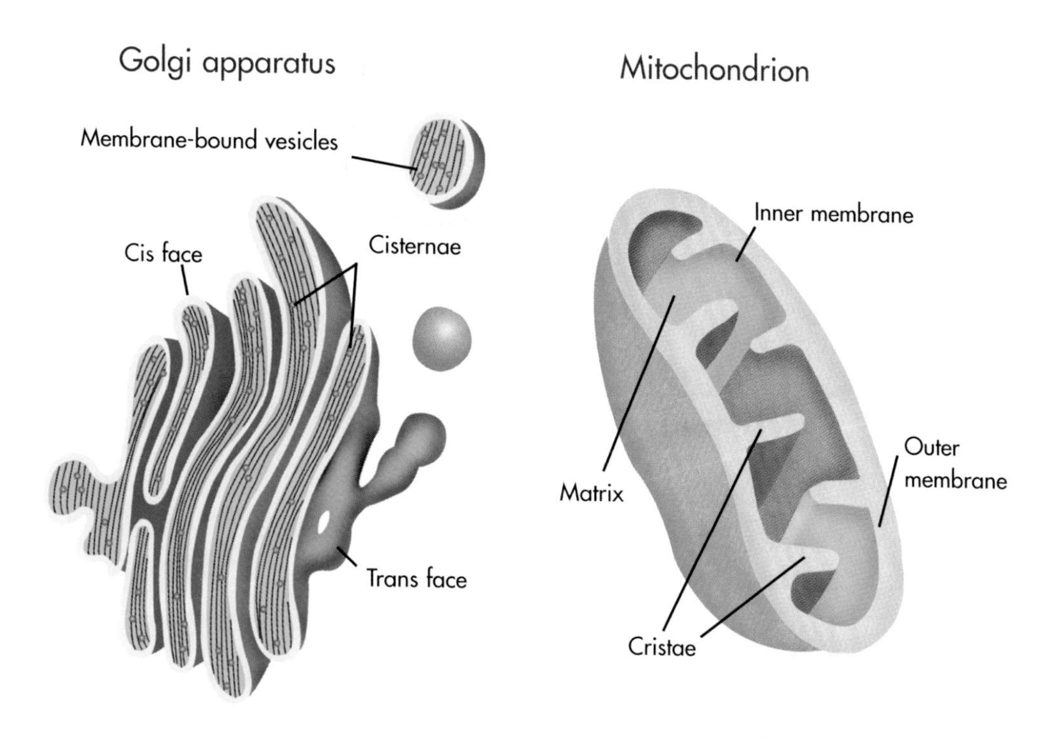

These two organelles, the Golgi apparatus and the mitochondrion, are central to cell function. The Golgi apparatus is a stack of flattened membrane-bound sacs called cisternae. The purpose of the Golgi apparatus is to receive and chemically change the molecules that were made in the endoplasmic reticulum, or ER. Its cis face is closest to the ER, while its trans face is on its opposite side. Mitochondria (singular: mitochondrion) are responsible for cellular respiration. They are bound by double membranes, contain their own DNA inside the matrix (the projections into the matrix are called cristae), and reproduce by dividing in two.

Like the nucleus, the mitochondria have a double membrane. The outer membrane is smooth, but the inner membrane has lots of wrinkles and folds, which increases its surface area. It is on the inner membrane that glucose reacts with oxygen to produce the primary energy for the cell.

Lysosomes are round organelles in the cytoplasm that contain enzymes. Lysosomes combat harmful substances and help digest food. Lysosomes are common in animal cells, but rare in plant cells. Normally, the lysosomes do not release the enzymes out into the cell itself. However, if the cell becomes damaged, the skins of the lysosomes disappear, enzymes are released into the entire cell, and the cell digests itself. This process of cell death is called necrosis.

Recent scientific research has helped scientists reach a new understanding of cells. Newly developed techniques in electron microscopy and light microscopy have revealed that cells contain a meshlike substance throughout the cytoplasm. This meshlike substance is called the cytoskeleton. It gives cells shape, anchors some organelles in place, and directs the movement of other organelles. The discovery of the cytoskeleton answered many unsolved mysteries about cells. Because of its many varied functions, the cytoskeleton is often referred to as both the "bones" and the "muscles" of cells.

Chapter Two

Plant Cells

Plant cells are much easier to see under a microscope than animal cells for two reasons. The first reason is that plant cells are generally larger than animal cells. In addition, plant cells have a thick cell wall outside the plasma membrane, making them easy to identify. Animal cells have only a thin, flexible cell membrane. The plant's thick cell wall is made of a rigid substance called cellulose, which helps the plant cell maintain its shape. Cellulose also protects the plant cell from mechanical damage. Cellulose is made up of linked glucose units, and it is the major raw material component used in the production of some manufactured fibers, such as rayon. Historically, paper was produced from the cellulose of the cell walls of cotton plants, which are 91 percent cellulose. Today, hardwoods and softwoods provide the major source of papermaking fibers, which are now made of only about 60 percent cellulose.

Plant cells are primarily composed of a large central vacuole, which fills at least 90 percent of

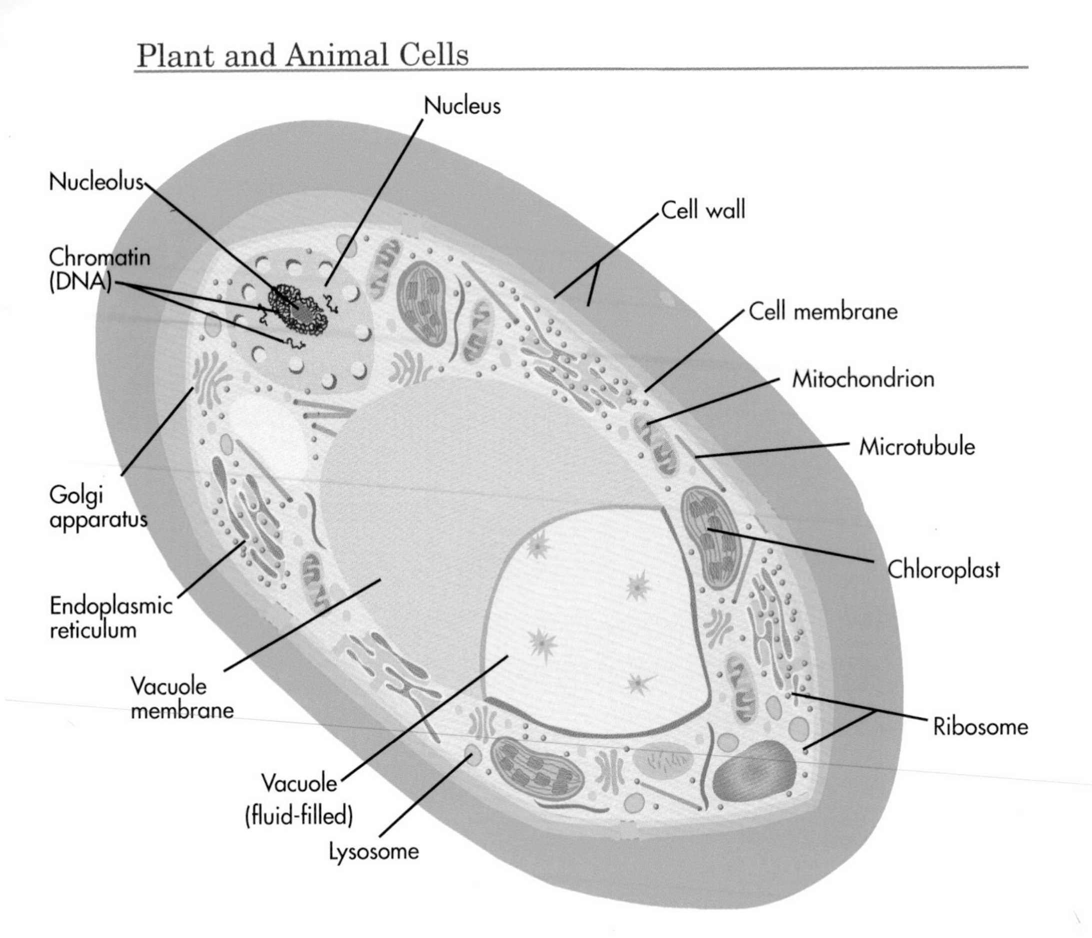

Components of a typical plant cell are shown in this illustration. Like eukaryotic animal cells, plant cells contain many of the same organelles including a nucleus, Golgi apparatus, mitochondria, and endoplasmic reticulum. Unlike animal cells, plant cells have a strong outer cell wall and chloroplasts, the large green organelles that trap sunlight and use it to make food.

a mature plant cell. The vacuole is a large watery bag near the center that stores chemicals and functions as a lysosome. Vacuoles can also help protect plants against predators by storing compounds that are poisonous to animals.

Cell Growth

The large central vacuole and the thick cell wall work together to aid the growth process for the entire plant.

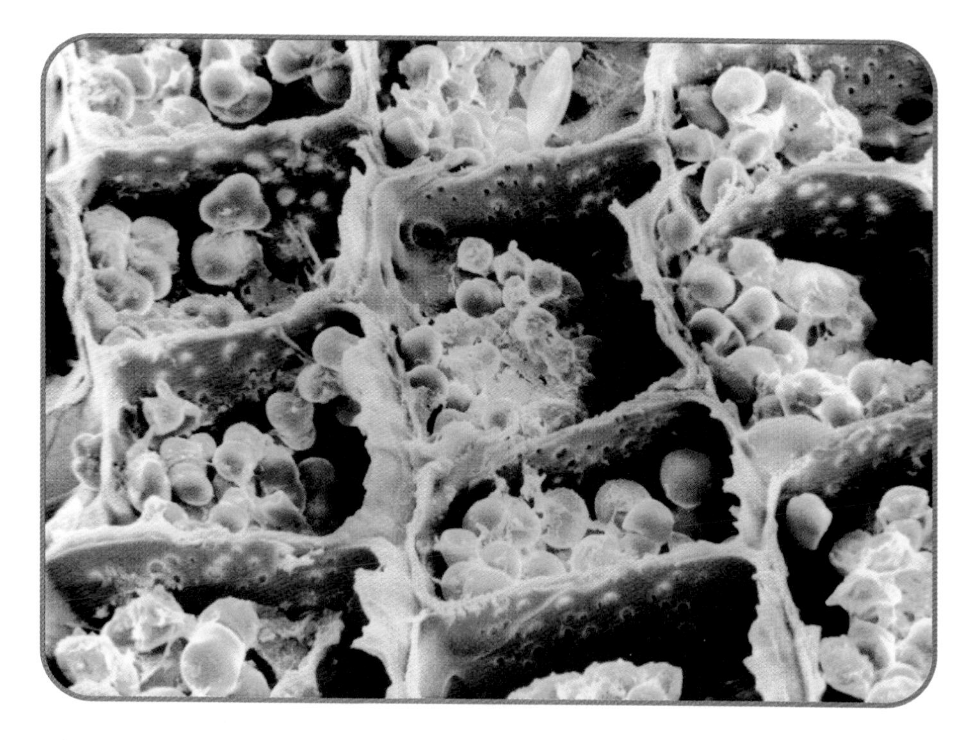

Above is a scanning electron micrograph (SEM) of cells from the genus *Mimosa* (tropical herbs, shrubs, and trees). Plants such as this one are supported by the rigidity of the cells when their cytoplasm and vacuoles are filled with water. The level of fluid inside the plant's cells that determines if the plant leaf or stem is rigid or limp is referred to as its level of turgidity.

Forget to Water Your Garden?

If a plant does not absorb enough water, it wilts. When you see plant leaves drooping, it is because the turgor pressure in its cells has dropped. When this happens, the cells can no longer support the plant. After a while, the plant will lose strength and undergo plasmolysis. In this process, the vacuoles of the plant cells shrivel and pull the cytoplasm away from the cell walls, causing the plant to die.

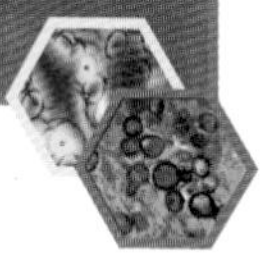

Most plant growth continues throughout its life due to cell division and cell enlargement. Plant growth occurs when the large central vacuole and the cytoplasm absorb water. By absorbing water, plant cells elongate, or grow longer. By contrast, animal cells grow by synthesizing organic molecules and increasing their cytoplasm. When the vacuole absorbs water in plant cells, it expands the contents of the cells within the cell walls, causing the plant to stiffen and have turgor pressure. If there is not enough water in the vacuole, the plant will collapse and wilt.

Transpiration

Transpiration is the cycle in which water passes through a plant. The outer cells of the leaves are constantly losing water through evaporation. This is especially true on a hot sunny day when water evaporates more quickly. When this occurs, the plant cell's vacuoles shrink in size because they are losing water. Therefore, the concentration of minerals and sugars in the outer vacuoles becomes higher than the concentration of minerals and sugars in the cell's inner vacuoles. The rate of evaporation is slower inside the cell's inner vacuoles because they are not directly under the sun's rays. Water then travels from the cells with a higher concentration of water to those on the surface with a low concentration of water. In doing so, water is pulled up through the stem and roots. This process of transpiration occurs constantly, though at different rates depending on how much water is in the soil.

Why Plants Are Green

Unlike animal cells, plant cells and certain types of algae contain chloroplasts, which are organelles that contain a green pigment called chlorophyll. All green parts of the plant contain chloroplasts including the stem, the leaves, and sometimes unripe fruit. The chloroplasts capture and trap light energy from the sun. This process is called photosynthesis. The chloroplasts then store this energy in a chemical form, and in turn, provide energy for all other living things in the form of glucose.

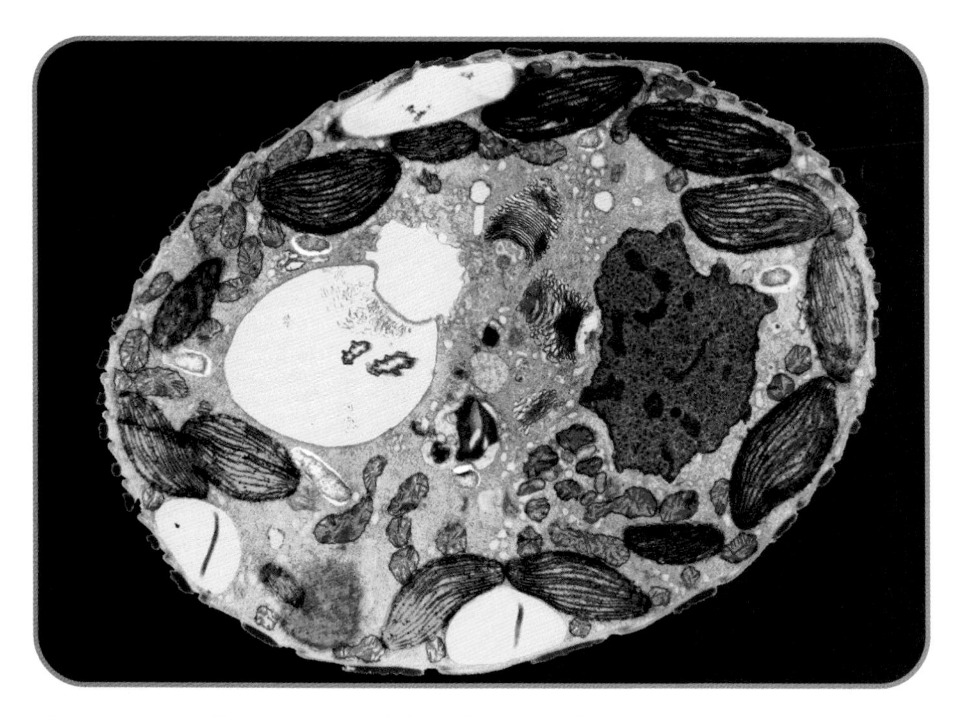

The chlorophyll-containing chloroplasts in this color-enhanced electron micrograph (TEM) are green. Other major organelles can also be seen such as the plant cell's nucleus, Golgi apparatus, mitochondria, endoplasmic reticulum, and vacuoles.

Chlorophyll

Chlorophyll, the green pigment inside the chloroplasts, absorbs red and blue light rays from the sun but reflects green light rays. This is why leaves are green. But in autumn, the chlorophyll in most plants breaks down, and other pigments such as carotene and tannin are more abundant. These other pigments reflect orange, yellow, and brown light making the leaves display a stunning array of fall colors.

The pigment in the chloroplasts, chlorophyll, traps the light energy. If you have ever worn a black shirt on a very sunny day, you know that dark colors absorb more light than light colors do. You most likely felt warmer than normal because the more pigment a substance has, the more light it will trap.

A plastid is a small particle in the cytoplasm that contains pigments. A chloroplast is a type of plastid, which contains the pigment chlorophyll. Chromoplasts are another type of plastid in plant cells that contain pigments other than chlorophyll, which give fruits, flowers, and autumn leaves their orange and yellow hues. Chromoplast pigments can be a variety of colors, but the chloroplast pigment is always green. Sometimes, chloroplasts will convert into chromoplasts. When this occurs in fruit, the color changes from green to a color characteristic of that fruit. This color change indicates to animals and humans that the fruit is ripe and ready to eat.

Well-known in desert regions, cactus plants like this one are often referred to as succulents. Cacti have adapted well to areas that have little rainfall. Cactus leaves have over time evolved into spines, the prickly, pointy projections seen on this plant. The cactus spines allow for less water to evaporate and help defend the plant from water-seeking predators.

Some plant cells contain more chloroplasts than others do. The amount of chloroplasts is determined by the position of the cells inside the plant. Since leaves are the main sites of photosynthesis, cells on the inside of the leaves contain the most chloroplasts. In most plants, the flat surface of the leaves provides a large area for sunlight absorption. But in cacti, the needles are "leaves." The chloroplasts of cacti are located in the green spongy stem rather than in the leaves because photosynthesis occurs in the stems of cacti rather than in the leaves.

Chapter Three

Animal Cells

Animal cells are harder to see under the microscope than are plant cells. Animal cells are often slightly smaller than plant cells, and they lack the thick cell walls that make plant cells unique.

The animal cell has an outer thin layer called the cell membrane or plasma membrane. Unlike the plant's rigid cell wall, the animal cell's plasma membrane is thin and flexible so that the cell can change its shape. The plasma membrane serves both as a protective coat and a gatekeeper that selects what to let in and out of the cell. The plasma membrane blocks the path of some substances, but lets others pass freely into and out of the cell. For example, the plasma membrane allows amino acids inside the cell for building proteins and carbohydrates inside for energy. The plasma membrane also regulates the amount of water inside the cell.

The plasma membrane's structure allows it to perform its function of being the gatekeeper of the cell. It is made up of two layers of lipid molecules. These lipid molecules hold proteins in place. The areas where the proteins lie mark the

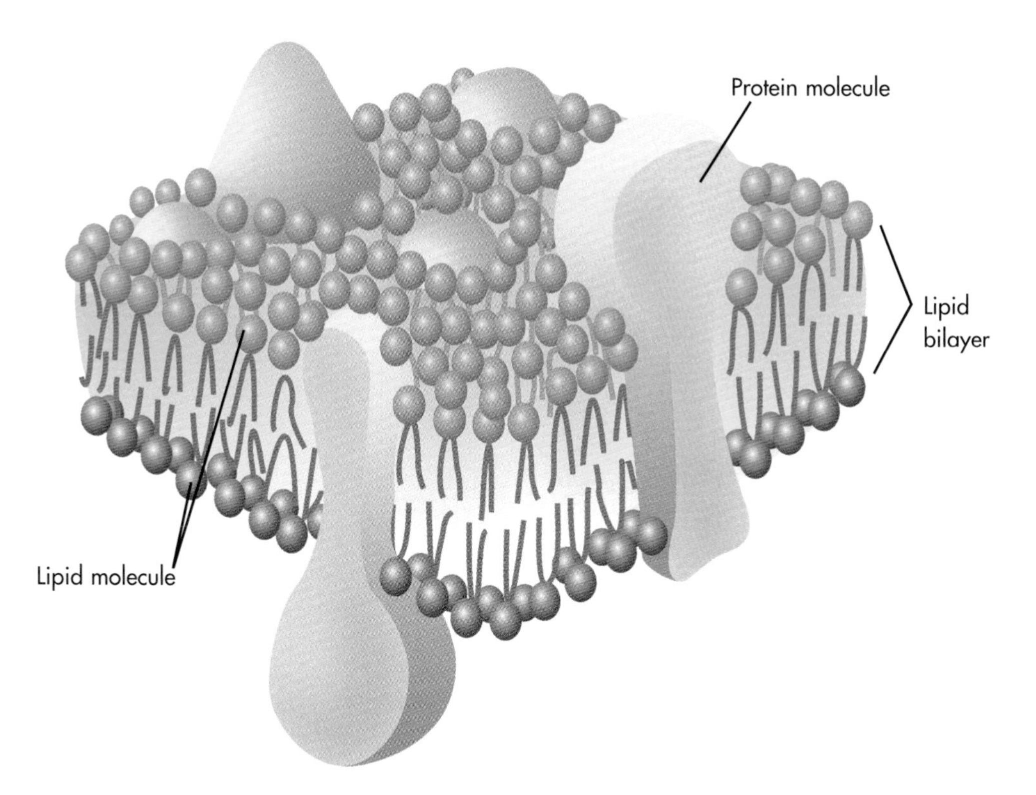

Cell membranes are made of lipids (fats) and proteins. The structure of a cell's membrane is a semipermeable two-layer system called the lipid bilayer, which is arranged from millions of lipid molecules. The lipid "head" is hydrophilic ("water loving") and its tail is hydrophobic ("water hating"). This system allows proteins to pass into the cell and wastes to pass out of it.

actual entrance and exit of molecules into the cell. Different proteins act as different entry and exit points for specific molecules.

In fact, similar membranes surround all of the cell's organelles. These membranes perform similar gatekeeper functions for other parts of the cell including the endoplasmic reticulum, the nucleus, the Golgi apparatus, and the mitochondrion. Membranes divide regions of the cell into different compartments. For example, the nucleus is wrapped in a double

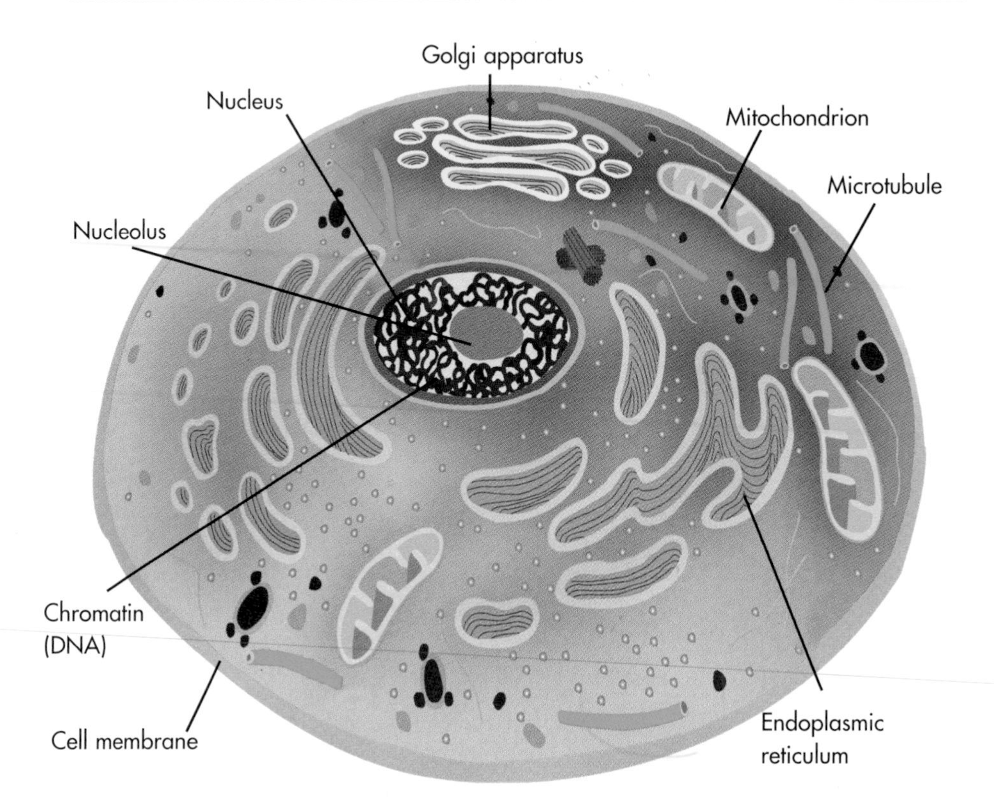

All organelles inside a eukaryotic animal cell are enclosed by single membranes, except for the nucleus and mitochondria, which are protected by double membranes. These membranes help the various organelles absorb materials such as proteins and enzymes, while helping them export manufactured products and waste.

membrane with pores or holes that let genetic information pass between it and the rest of the cell.

So far, you have read about characteristics unique to all plant cells and characteristics unique to all animal cells. But each plant and animal, as a whole, contains many different types of cells in a variety of shapes and sizes. This variety of cells allows the cells to do their special jobs and contribute to keeping the whole organism alive and functioning properly. Groups of similar cells that work together to carry out a particular function are called tissues. For example, animals have muscle tissue, nerve tissue, and bone tissue. Plants have vascular tissue, xylem tissue, and phloem tissue.

Muscle Cells

A group of muscle cells together form muscle tissue. Individual muscle cells are long and thin and can shorten or lengthen (contract and release) to allow for movement. You know how tired you sometimes feel after running? That is because your muscles use up a lot of energy every time they have to contract

Smooth human muscle tissue, like that found in the walls of the stomach, intestine, uterus, bladder, and arteries, is pictured in this electron micrograph. Smooth muscle cells are thin and elongated and form layers of tissues in sheets. Because smooth muscle tissue is controlled involuntarily, it doesn't contain the channels or grooves often found in skeletal muscle tissue.

and release. Therefore, each muscle cell has a lot of mitochondria, the organelles inside cells that serve as the energy producers of the cell.

The mitochondria provide a site in the cell for oxygen to combine with glucose. This combination produces the energy needed for the cell to carry out its functions. A person who moves a lot, such as an athlete, appears to have larger muscles. What is actually happening to his or her muscles is that the muscle cells are getting bigger. The athlete's body self-regulates to make sure that its muscle tissue will have enough mitochondria to keep up with his or her increased movement. Conversely, a body that remains idle experiences the reverse process, and his or her muscles shrink from nonuse.

Nerve Cells

Nerve cells, also called neurons, are very long and thin. They are wirelike extensions, along which nerve signals pass from one part of the body to another. Neurons are thinner than thread, and some can be as long as 3 feet (0.91 meters).

Much like telephone wires that are connected in a network and transmit messages, neurons also carry messages over a network throughout a human's or an animal's body. The center of this network is the brain and the spinal cord.

Both humans and animals have sensory neurons and motor neurons. Your sensory neuron cells allow you to feel sensations such as pressure, heat, cold, or pain by carrying messages from such places as your eyes or your ears to your brain. All animals' bodies send and receive messages in this same way.

Just as in animals' bodies, your motor neurons carry messages from the brain and spinal cord to other parts of the body to give the message to react by moving specific muscles.

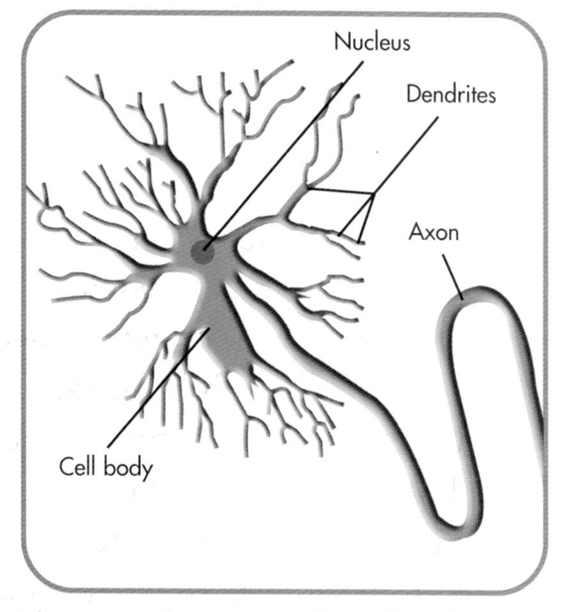

The parts of a nerve cell—cell body, dendrites, nucleus, and axon—are represented in this illustration. The human brain contains about a hundred billion nerve cells, or neurons, each one of them rapidly firing to produce sensory information. The neuron's axon acts as an electrical "conductor," sending information between cells, while the dendrites receive that information. Like all other cells, the nucleus controls the cell and contains its genetic material.

Interneurons connect sensory neurons to motor neurons. If you accidentally touch a hot stove, all three types of neurons will work together to immediately send a message to your hand, "Remove from hot stove now!" Nerve cells are connected to each other in long chains by branchlike dendrites at one end of each nerve cell and a long axon at the other end. The axon transmits the message to the neighboring cell's dendrites. The chains of nerve cells carry messages from nerve cell to nerve cell, until it reaches its appropriate destination.

Bone Cells

Bone cells have concentric circles, which means they have many circles within circles. They look like the concentric circles you see when you slice an onion. Bone cells make a liquid that hardens like cement that allows the bone to have a strong structure. There are many different types of bone cells.

Although a bone is very hard, it is actually hollow inside. It is filled with bone marrow, a combination of red and yellow marrow cells. Bone marrow is the factory of blood cells. The red bone marrow cells make red and white blood cells. Yellow bone marrow produces some white blood cells, but it also contains fats and connective tissue. Some bone cells make new bone cells. These bone-building cells allow your bones to repair themselves if they break.

Antler Bones

A deer's antlers, made of bone, are one of the fastest-forming tissues known. Although bucks (male deer) shed their antlers every spring, they start growing back immediately. Antlers can grow up to a quarter-inch (0.64 centimeters) per day. Compare that to human hair, which grows approximately one-half an inch (1.3 cm) per month.

Other bone cells actually consume bone from the inside. These self-consuming bone cells ensure that the interior of your bones remain hollow so that bone marrow has room to grow. As a human or animal body grows from childhood to adulthood, its bones must grow, too. Therefore, the interior of your bones must get bigger in order to make more space. Thanks to the bone-eating cells, your bones continuously make room for the bone marrow that makes your blood cells, even as your body grows bigger.

Epithelium Cells

Epithelial tissue can be found on both the outside of an animal's or a human's body and the lining of the organs. Epithelial cells can be found in an animal's intestines, lungs, skin, and blood vessels. In fact, all the organs in both a human's and an animal's body are covered with epithelium cells. They come in all different shapes. Some epithelial cells are flat, and others are cubes or columns.

The epithelial cells lining the insides of the intestines allow digested food to pass through and into the blood. The lung's many air sacs, called alveoli, are lined with epithelial cells, too. The lining of the alveoli is extremely thin so that gases, such as oxygen, can pass freely through it and go to your cells for cellular respiration.

Your whole body is wrapped in a covering of epithelial cells, just like an animal's body. In a

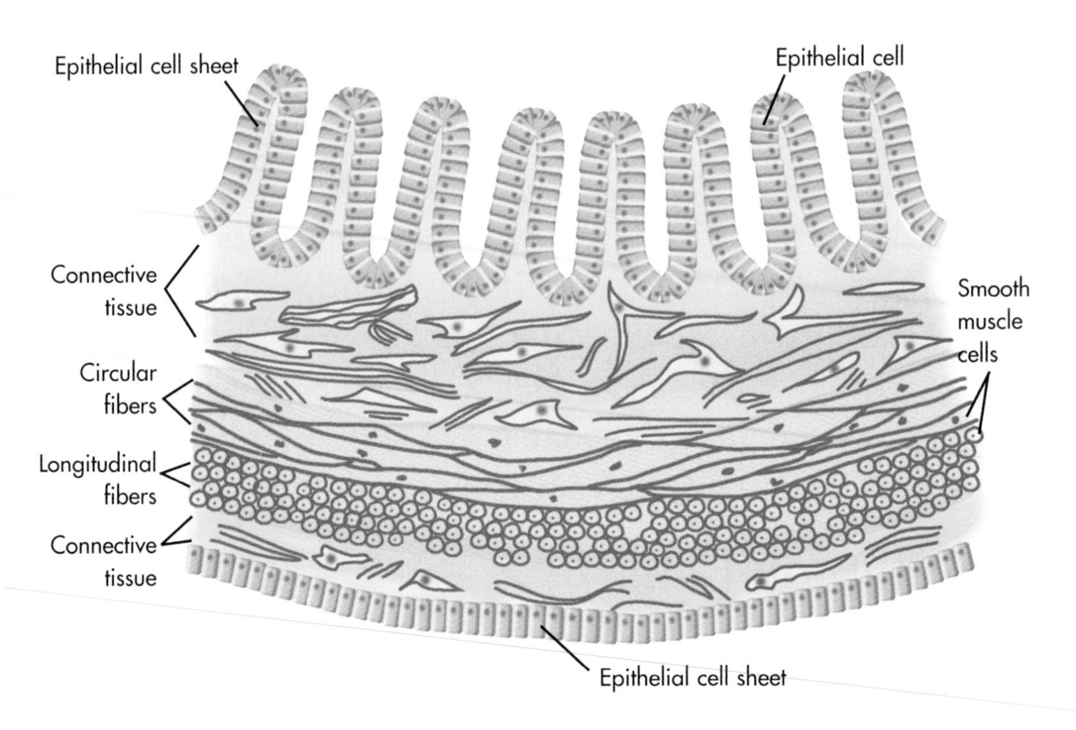

This cross-sectional diagram of a mammal's intestine helps illustrate how cells are formed together to create various types of living tissue. The long tubular intestine is constructed of epithelial cells (sepia-toned), while connective tissue is colored blue. Muscle tissues, both smooth cells and circular fibers, are colored yellow.

human being, the outer layer of skin, called the epidermis, is made up of epithelial cells, which are flat, hard, and tough. (The underlying skin layer is called the dermis.) Skin cells protect the body's interior parts from germs. If you cut yourself, however, the germs can get inside your body. If they do, you could get an infection.

You lose epithelial cells from your skin every minute of every day. Every time you wash your hands or rub your skin, you lose some epithelial skin cells. However, they regenerate themselves so

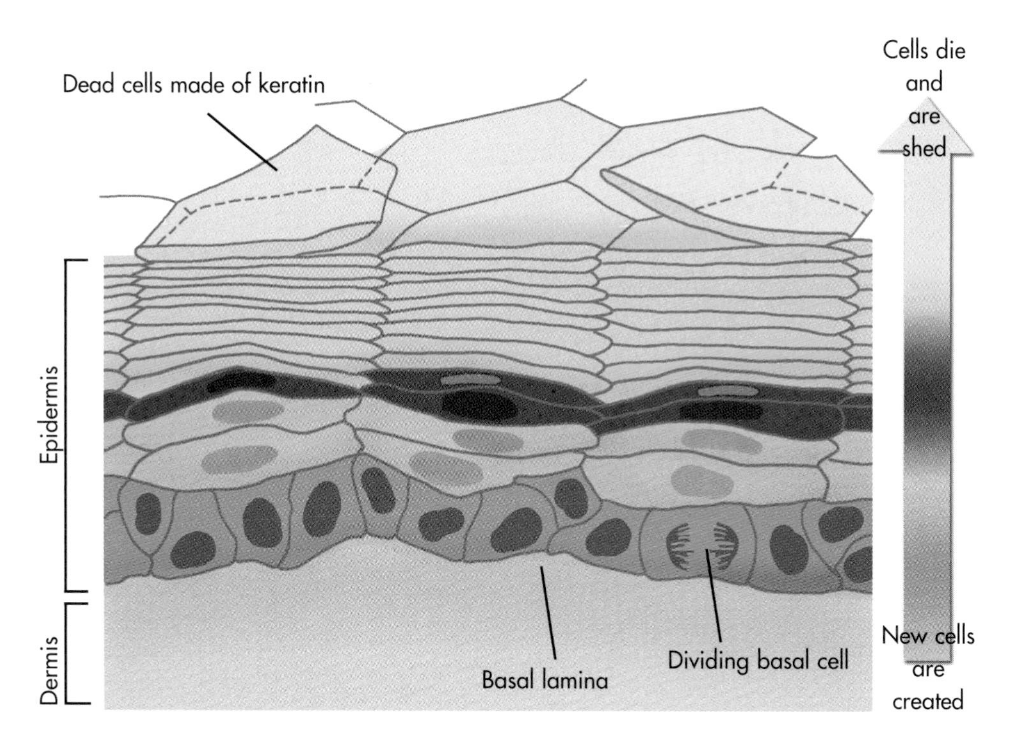

In this illustration of human skin layers, the shedding process of cells is shown as flattened cells on the surface die. Skin cells are first created in the basal layer and move outward as they differentiate and age. At the time of cell death, the cell's nucleus disintegrates, and it flattens and shrinks until it is only a flaking scale made up of keratin, a fibrous protein.

you never even notice. In fact, in a year's time you lose as much as 5 pounds (2,268 grams) of dead skin cells.

The insides of blood vessels are lined with a type of epithelial cells called squamous epithelial cells. Squamous cells have the appearance of thin, flat plates. The shape of these cells allows the inside of the blood vessels to be very smooth so as to allow blood to flow easily without the blood cells becoming damaged.

Chapter Four

Plant Cell Functions

Plants have vascular tissue running up and down their stems. Most plants are vascular plants, meaning they have leaves, roots, and stems. These leaves, roots, and stems work together to absorb sunlight, anchor the plant, and transport food and water. Plants that are nonvascular include mosses and lichens. Nonvascular plants perform all of these same functions without true stems, roots, leaves, or vascular tissues.

Vascular tissue carries fluids and helps support the plant. If you have ever seen a cross section of a tree stump, you can easily see the vascular tissue that makes up the transport system of vascular plants.

Vascular tissue is made up of both xylem tissue and phloem tissue. Xylem tissue carries water and minerals up through a plant from the roots to its branches and leaves. Phloem tissue distributes food made in the leaves to all parts of the plant. Both of these tissues can be found in plant stems. The stem, therefore, serves as the transportation system for the plant. The stem is like a two-way highway from the soil to the surface of the leaves.

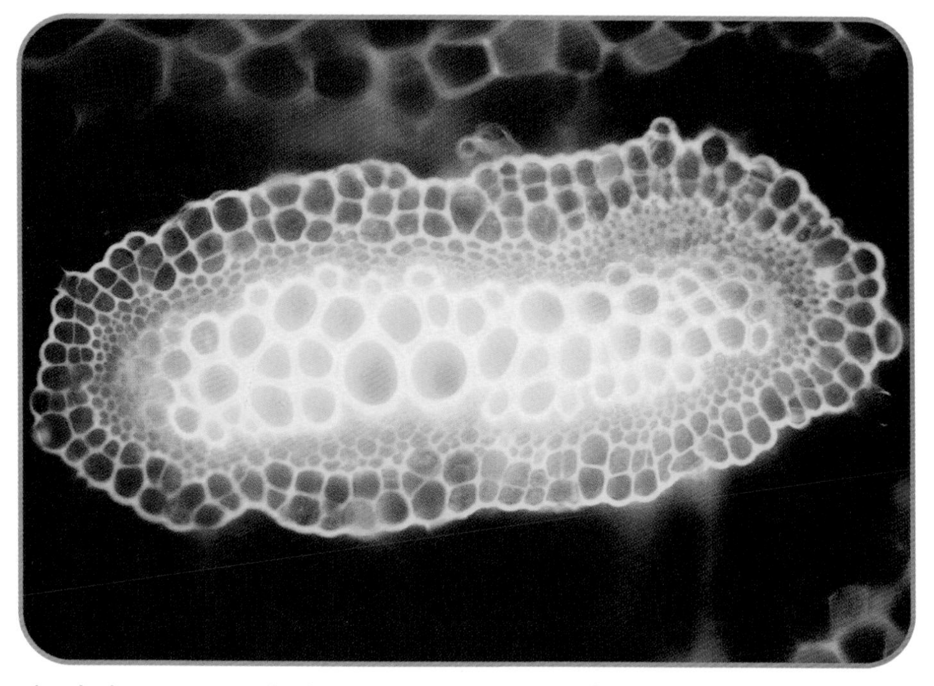

This light micrograph shows a cross section of an evergreen mosslike plant from the genus *Selaginella*. Note its arrangement of vacuoles, which, when filled with water, provide a rigid support system. Plants of this species can thrive almost anywhere in the world but are mostly found in tropical climates.

Xylem cells are long, hardwalled, and tubeshaped, and they join from end to end. When they die, their contents dissolve. The cells then become microscopic tubes through which water flows to reach the different parts of the plant. Although plants do not have bones, their xylem keeps them firm. The wood we get from tree trunks is made up of xylem tissue. The stalk of a flower and even a blade of grass also contains xylem, which helps keep them upright. In these smaller plants, water also plays a large role in keeping the stalks stiff and tall. In small plants, turgor pressure, which is created when the large

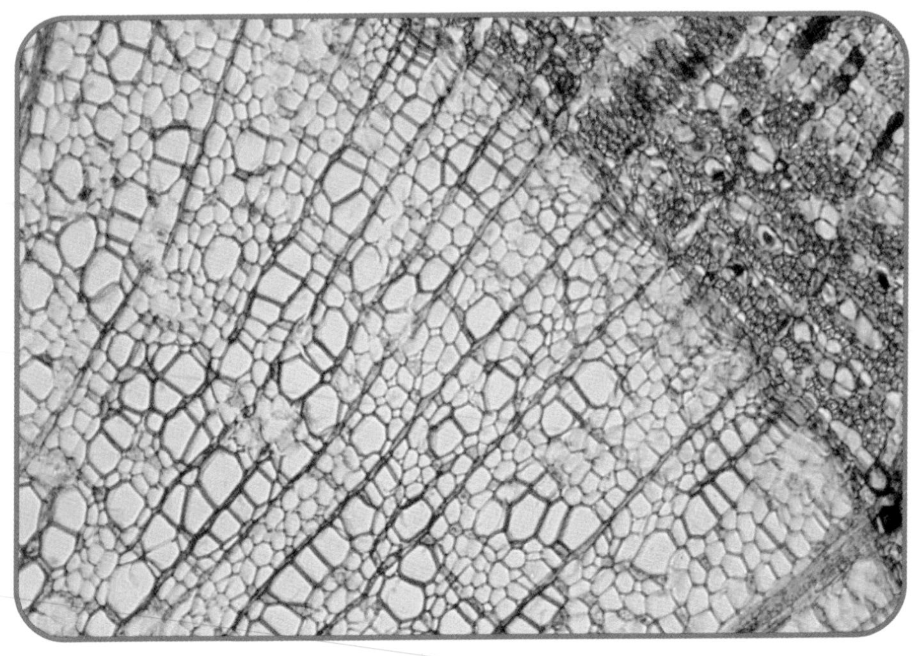

This two-year-old stem of a lime tree, as seen in a microscopic cross section, shows the tree's xylem *(bottom left)*, cambium layer (narrow, curved, green band) and phloem *(upper right)*. The xylem is composed of vertical, tubelike cells that conduct the sap upward from the tree's roots. The phloem is responsible for the distribution of the nutrients produced by photosynthesis in the tree's leaves.

central vacuoles of the plant cells absorb water, also keeps the plant erect. Both the xylem and the turgor pressure work together to keep the plant upright.

Phloem cells form long tubes for conveying food-rich sap manufactured in the leaves to other parts of the plant. Pith cells are a central area of tissue found in stems, and the pith tissue acts as a filling in the stem. Pith tissue is also used for storing food.

Epidermis—Nonwoody Stems

The outside of a green leaf or stem is covered with a layer of cells called the epidermis. The cells of this outer layer make a waxy substance that acts as a smooth, waterproof coating for the plant. This coating can only be found on herbaceous, nonwoody stems. The coating allows the stem to fill up with water, making it stiff. The epidermis consists mainly of parenchyma (packing tissue) and collenchyma (supporting tissue).

Bark—Woody Stems

Nonherbaceous stems have an outer protective layer of tough bark. After many years, the woody plants form new vascular tissue through secondary thickening. The results are a woody plant or a tree. Year by year, new vascular tissue is formed, increasing the diameter of the tree trunk. If you have ever looked at a tree trunk, you have probably seen a number of concentric circles. Each year new layers of xylem are produced. Large xylem cells are produced in the spring and are lighter

This cross section of a laburnum tree shows the circular bands of growth that took place over the course of the tree's life. Counting these rings reveals the age of the tree at the time of its felling. Botanists can learn a number of things by studying tree rings, including environmental conditions that might have affected the speed or amount of growth from year to year.

in color. The production rate of new cells slows during summer, producing smaller, darker xylem cells. The changes in the size of xylem cells generated during varying seasons produces annual rings, concentric circles that are a good indicator of the age of a tree.

Guard Cells

A leaf consists of a thin, flat blade, supported by a network of veins, a leaf stalk called a petiole, and a leaf base where the leaf joins the stem. Stomata are tiny openings in the epidermis through which the exchange of water and gases takes place. They are mainly found on the underside of the leaves.

There are thousands of guard cells on the underside of each leaf. The guard cells come in pairs and look like two tiny kidney beans, placed side by side so that there is a small hole in the middle.

The structure of the guard cells allows them to perform their special function to keep the plant alive. The guard cells open and close depending on whether they are letting in water, sunlight, or gases from the air. It is through these openings that the plant "breathes."

Root Cells

Much like the epithelial cells that line the intestines of animals and aid in digestion, the root cells of plants also help it absorb nutrients.

Roots, the underground parts of plants, anchor them in soil and help them absorb water and minerals from that soil. The roots' absorptive properties are increased by root hairs, which grow behind the root tip, allowing the maximum amount of surface for absorption.

A carrot is the root of a carrot plant. If you were to see a carrot growing in the ground, all you would see is a bunch of green leaves with thin stems above the dirt. To find the carrot, you have to pull the leaves and pull up the big root from underground. When you look at round carrot slices in a salad, you can see a center core that is slightly darker than the outer core. These two sections contain different kinds of cells. Roots have an outer epidermal layer, a cortex of parenchyma packing tissue, and a central cylinder of vascular tissue.

Meristem cells, located near the tip of the root, divide to produce new cells. This elongation of the cells pushes the root farther down into the soil. A layer of cells called the root cap, which covers the tip of the root, protects the root tip as it grows.

Chapter Five

Photosynthesis and Cellular Respiration

Plants are Earth's natural solar panels. Because their cells contain chloroplasts, plants capture the energy of the sunlight, use its energy to break down atoms, and in turn, provide energy for all other living things in the food chain.

Capturing sunlight and producing energy for all living things is a very simple way of describing a complicated chemical reaction that occurs during photosynthesis. In this reaction, carbon dioxide combines with water, and the atoms are broken down by the light energy captured from the sun to form carbohydrates and oxygen. Carbohydrates are the energy-containing result of the chloroplasts capturing energy from the sun. The chemical equation for the photosynthesis reaction looks like this:

$$6CO_2 + 6H_2O + \text{sunlight energy} \dashrightarrow C_6H_{12}O_6 + 6O_2$$

Don't panic. It's a simple formula. CO_2 stands for carbon dioxide. A carbon dioxide molecule has one atom of carbon and two atoms of oxygen. A water molecule is referred to as H_2O

because it is made up of two atoms of hydrogen and one atom of oxygen. This means that each individual reaction requires six carbon dioxide molecules, six water molecules, and some sunlight energy to produce six molecules of oxygen and one molecule of glucose (a carbohydrate).

Carbon dioxide enters the leaf, and oxygen exits, by way of microscopic pores called stomata flanked by two guard cells. The carbon dioxide combines with the oxygen and hydrogen in water stored in the leaves to form organic material called carbohydrates plus oxygen. Oxygen exits the plant through the stomata in the leaves as a waste product of photosynthesis. Animals breathe in oxygen, which is provided by plants, and breathe out carbon dioxide.

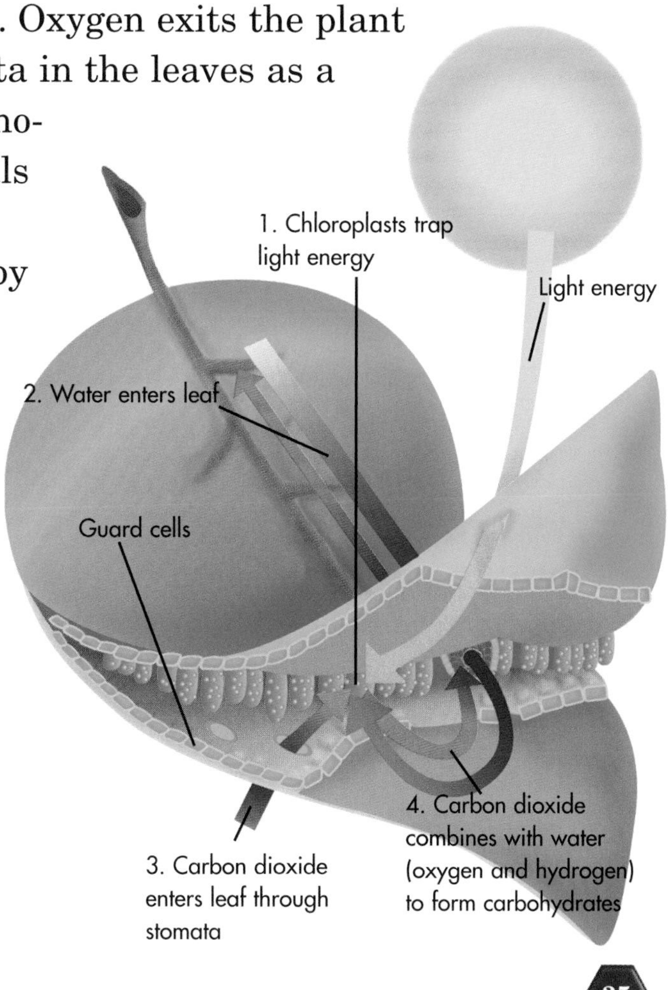

Light-absorbing chlorophyll, contained in the green leaves of all plants, absorbs sunlight in order to make nutrients. This chemical reaction, which is illustrated in this diagram, is called photosynthesis. Plants release oxygen as a waste product during this process.

The process of photosynthesis allows plants to be self-fed, or autotrophic. Autotrophs sustain themselves without eating other organisms as humans and other animals do. As autotrophs, plants derive their nutrients from the sun's light energy, carbon dioxide in the air, minerals in the soil, and water.

Chloroplasts are the structures in cells that capture and trap light energy from the sun during the process of photosynthesis. The pigment inside the chloroplasts, chlorophyll, is actually the substance that traps the light energy.

The leaves of plants are where photosynthesis takes place. The flat surface of the leaves provides a large surface area for sunlight absorption, and similarly, the cells of leaves contain many chloroplasts. An extensive network of veins called xylem tissue brings water into the leaves. Phloem tissue transports glucose (carbohydrates) produced by photosynthesis from the leaves to the rest of the plant.

Plant cells' ability to undergo photosynthesis has caused some scientists to hypothesize that the solution to global warming may be to plant more trees. Global warming is an ecological crisis caused by unusually high levels of carbon dioxide and other greenhouse gases in the atmosphere. Greenhouse gases cause heat to be trapped near the surface of Earth, which may have raised its temperature over the past century. Some scientists say that having more forests on Earth means that more plant cells will undergo photosynthesis, translating into lower levels of carbon dioxide in the atmosphere. This theory remains a subject of heated debate among scientists.

Cellular Respiration

Animals could not undergo cellular respiration without the glucose and the oxygen produced by plants during photosynthesis. Animals ingest glucose and mitochondria in their cells transform the glucose into energy. During cellular respiration, the mitochondria break down glucose into atoms, and by

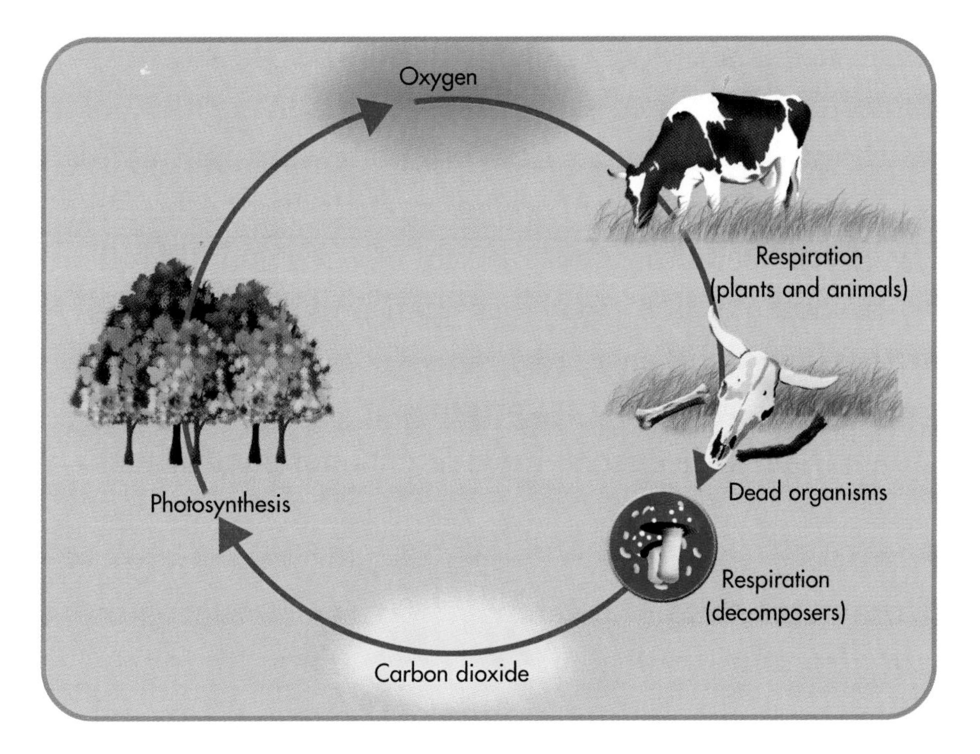

All living things exchange gases within the environment in a process called cellular respiration. Cellular respiration and photosynthesis are interdependent, as noted in this diagram. While plants give off oxygen as a waste product of photosynthesis, humans breathe in oxygen and exhale carbon dioxide as a waste product. Carbon dioxide is then absorbed by plants through openings in their leaves called stomata in this cycle.

mixing it with the atoms from oxygen, produce carbon dioxide, water, and energy. By taking in food molecules and oxidizing (burning) them, mitochondria provide energy for the cell.

Cellular respiration of animals is the exact opposite gas exchange process than is photosynthesis. The equation for the photosynthesis of plant cells is $6CO_2 + 6H_2O$ + sunlight --> $C_6H_{12}O_6 + 6O_2$, while the equation for cellular respiration of animal cells is: $6O_2 + C_6H_{12}O_6$ --> $6H_2O + 6CO_2$ + energy. According to the equation, in each individual reaction, the mitochondrion needs six oxygen molecules and a glucose molecule to produce six carbon dioxide molecules, six water molecules, and a new form of energy.

These two systems depend on each other. The ability of your cells to respire is what keeps you alive. The by-products of photosynthesis can provide organisms with oxygen and carbohydrates necessary for cellular respiration, while the by-product of cellular respiration can provide carbon dioxide for photosynthesis.

Oxygen and the Circulatory System

Organisms take in oxygen either from water or the air, depending on the organism. Humans take in oxygen when inhaling; fish take in oxygen through water. The oxygen combines with carbohydrates from digested food, then organisms release carbon dioxide to the environment when exhaling.

Oxygen is spread around the body by the circulatory system. All the cells in the body require a constant supply of oxygen. Oxygen is carried to the cells by the bloodstream and carbon dioxide is carried away. The circulatory system is made up of tiny branching blood vessels for this purpose. Like a circle, the system has no beginning or end. It takes one minute for your blood to make a complete loop around the circulatory system.

Mammalian lungs have a huge surface area to receive the maximum amount of oxygen possible from each inhalation. The lungs have a branchlike air tube, the bronchi, and bronchioles, which allow for the most exposure to oxygen. The bronchioles

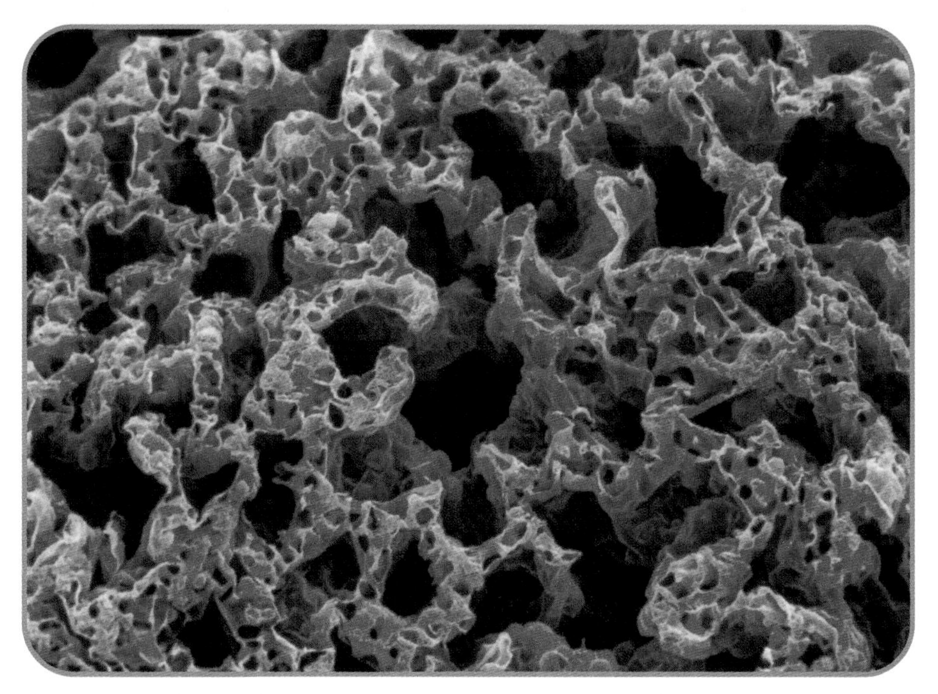

This color-enhanced scanning electron micrograph of human lung tissue shows alveoli, the tiniest capillaries of the human "bronchial tree." Blood is carried to the alveoli where it receives oxygen that it then carries throughout the entire body. The average pair of human lungs contains more than 600 million alveoli.

end in microscopic air sacs called alveoli, which have a moist coating that dissolves oxygen. A layer of epithelium tissue lets the oxygen pass through to the capillaries through which the oxygen is carried by blood flowing through vessels to every cell of the body.

Fish take in oxygen when water passes through their gills. The oxygen passes easily through the very thin, delicate skin of the gill filaments, into the blood flowing just beneath them.

The red blood cells carry oxygen through the body, drop it off at other cells, then pick up carbon dioxide, and carry the carbon dioxide back to the lungs. Mammals then breathe out the carbon dioxide into the air. As with all cells, the red blood cells' shape helps them perform their function. The red blood cells are disk shaped so that they can carry oxygen to, or carry carbon dioxide away, from cells.

Plant and Animal Cells Are Interdependent

Plants and animals must coexist in order for either to flourish. The interdependence of the organisms is based on the interdependence of their cells. This fundamental interdependence comes from the need of every cell to survive. In addition to the relationship of photosynthesis and cellular respiration, the interdependence between plants and animals can be found in many other ways, too, such as when bees help pollinate flowers while they get nectar, or when animals help disperse seeds by catching them in their fur or paws.

The interdependence of plants and animals is the basis for the branch of science called ecology. By learning about cells, the most basic units of life, we can begin to understand larger biological systems as a whole.

Glossary

alveoli (ahl-VEE-oh-lee) Microscopic air sacs in the lungs lined with epithelial cells.

cell (SEL) A small unit of living matter.

cell membrane (SEL MEM-brayn) A thin, semipermeable double layer of fatty molecules that surrounds every cell.

cell wall (SEL WAHL) A rigid wall outside the cell membrane of plant cells composed mostly of cellulose; it provides support and protection.

cellulose (SEL-yuh-lohs) A tough carbohydrate that makes up the thick cell wall of plants.

chlorophyll (KLOR-uh-fil) The pigment in green plants that absorbs solar energy.

chloroplast (KLOR-uh-plast) A plastid in plant cells that contains the pigment chlorophyll; provides the site for photosynthesis.

chromoplast (KROHM-uh-plast) An organelle in plant cells that contains pigments other than chlorophyll.

cytoplasm (SY-toh-plah-zum) A jellylike material that fills all the spaces in the cell between the plasma membrane and the nucleus.

cytoskeleton (SY-toh-skel-uh-tin) A meshlike network in the cytoplasm that provides internal support for the cells and anchors internal cell structures.

endoplasmic reticulum (EN-doh-plahs-mik reh-TIH-kyoo-lum) Long stringy structure in eukaryotic cells where proteins are synthesized and used for growth and repair.

epithelial cells (eh-pih-THEE-lee-yul SELZ) Cells in animals that cover both internal and external surfaces of the animal body with epithelial tissue.

eukaryote (YOO-kah-ree-oht) An organism with cells having a distinct nucleus and intracellular membranes. All protists, fungi, plants, and animals are eukaryotes.

Golgi apparatus (GOHL-jee ah-pahr-AH-tus) An organelle that handles protein processing and transport.

lysosome (LY-so-zohm) An organelle full of enzymes that enables a cell to break down certain molecules for recycling and disposal.

meristem cell (MEH-rih-stem SEL) A cell located near the tip of a root, which divides to produce new cells.

mitochondrion (my-toh-KON-dree-on) An organelle that serves as the site for the reaction of oxygen and glucose to produce energy used for the cell.

neuron (NUR-on) Very long and thin wirelike extension along which nerve signals pass from one part of the body to another; nerve cell.

nucleoli (noo-klee-OH-ly) A round body inside the nucleus that produces messengers that are sent out of the nucleus to other parts of the cell.

nucleus (NOO-klee-us) A part of the cell containing DNA and RNA, responsible for growth and reproduction.

organelles (OR-gen-ellz) Structures throughout the cell and within the cytoplasm that carry out a particular function in the life of the cell.

phloem (FLOH-um) Cells of the vascular system in plants that transport food from leaves to other areas of the plant.

photosynthesis (foh-toh-SIN-thuh-sis) Chemical reaction that occurs in the chloroplasts of plant cells in which plants capture the energy of the sunlight, use the energy to break down atoms, and in turn, provide energy for all other living things in the food chain.

plasmolysis (plaz-moh-LY-sis) Condition in which a cell loses water to its outside environment.

plastid (plah-STID) A small particle in the cytoplasm that contains pigments.

prokaryote (PRO-kahr-ee-oht) An organism with cells possessing no distinct nucleus, such as bacteria and cyanobacteria.

respiration (reh-spuh-RAY-shun) Process by which a mitochondrion breaks down glucose into atoms and mixes it with the atoms from oxygen to produce carbon dioxide, water, and energy for the cell.

ribosome (RY-boh-zohm) A cell organelle, usually found on the surface of the endoplasmic reticulum, which functions in the synthesis of proteins.

root cap (ROOT KAP) Layer of cells covering the tip of a root, which protects the root tip as it grows down into the soil.

tissue (TIH-shoo) A group of similar cells that work together to carry out a specific function.

transpiration (tranz-puh-RAY-shun) The cycle in which water passes through a plant.

turgor pressure (TUR-gur PREH-shur) Pressure caused by the cytoplasm pressing against the cell wall.

vacuole (VAH-kyoo-ohl) A large sac in the inside of plant cells that removes waste products and stores ingested food.

vascular tissue (VAHS-kyoo-lur TIH-shoo) Xylem and phloem tissue found in vascular plants.

xylem (zy-LUM) Tissue in the vascular system of plants that moves water and dissolved nutrients from the roots to the leaves.

For More Information

Discover Magazine
114 Fifth Avenue
New York, NY 10011
(212) 633-4400
Web site: http://www.discover.com

National Science Foundation
4201 Wilson Boulevard
Arlington, VA 22230
(703) 292-5111
Web site: http://www.nsf.gov

Web Sites

Due to the changing nature of Internet links, the Rosen Publishing Group, Inc., has developed an online list of Web sites related to the subject of this book. This site is updated regularly. Please use this link to access the list:

http://www.rosenlinks.com/lce/plac

For Further Reading

Maton, Anthea. *Cells: Building Blocks of Life*. Upper Saddle River, NJ: Pearson Prentice Hall, 1997.
Scott, Michael. *The Young Oxford Book of Ecology.* Oxford, England: Oxford University Press, 1995.
Wallace, Holly, and Anita Ganeri. *Cells and Systems* (Life Processes). Chicago: Heinemann Library, 2001.

Bibliography

Burnie, David. *Dictionary of Nature.* New York: Dorling Kindersley, Inc., 1994.
Campbell, Neil A. *Biology.* Upper Saddle River, NJ: Pearson Benjamin Cummings, 1990.
Eyewitness Visual Dictionaries. *The Visual Dictionary of Plants.* New York: Dorling Kindersley, Inc., 1992.
Parker, Steve. *Animal Biology.* New York: Prentice Hall, 1992.
Parker, Steve. *How Nature Works.* New York: Random House, 1993.
Silverstein, Alvin. *Cells: Building Blocks of Life.* Princeton, NJ: Prentice-Hall, Inc., 1969.
Stockley, Corinne. *The Usborne Illustrated Dictionary of Biology.* London, England: Usborne Publishing, Ltd., 1986.
VanCleave, Janice. *The Human Body for Every Kid.* New York: John Wiley & Sons, Inc., 1995.

Index

A

animal cells, 18–27
  dependence on plant cells, 40–41
  parts of, 18–19

B

bark, 31
bone cells, 21, 24–25
bone marrow, 24, 25

C

cells, parts of, 6–10
cellular respiration, 8–9, 25, 37–38, 40
cellulose, 11
chlorophyll, 15–16, 36
chloroplasts, 15–17, 34, 36
chromoplasts, 16
cytoplasm, 8, 10, 13, 14, 16

D

DNA (deoxyribonucleic acid), 6–7

E

endoplasmic reticulum, 8, 19
epidermis
  human, 26
  plant, 30–31, 32, 33
epithelial cells, 25–27, 32, 40
eukaryotic cells, 6

G

global warming, 36–37
Golgi apparatus, 8, 19
guard cells, 32, 35

L

lysosomes, 10, 12

M

meristem cells, 33
mitochondria, 8–10, 19, 22, 37–38
multicellular organisms, 5
muscle cells, 21–22

N

nerve cells, 21, 22–24
nucleolus, 7–8
nucleus, 6–8, 10, 19–20

O

oxygen, need for, 38–40

P

phloem tissue, 21, 28, 30, 36
photosynthesis, 15, 17, 34–38, 40
pith cells, 30
plant cells, 11–17
  cell growth and, 12–14
  dependence on animal cells, 40–41
  functions of, 28–33
  parts of, 11–12
plasma/cellular membrane, 6, 8, 11, 18

plasmolysis, 13
plastid, 16
prokaryotic cells, 6

R
ribosomes, 7–8
root cells, 32–33

T
tissues, 21
transpiration, 14
turgor pressure, 14, 29, 30

U
unicellular organisms, 4–5

V
vacuole, 11–12, 13, 14, 30
vascular plants, 28
vascular tissue, 21, 28, 31, 33

X
xylem tissue, 21, 28, 29, 30, 31–32, 36

About the Author

Judy Yablonski received a degree in environment science from Barnard College of Columbia University. She is both a freelance writer and a lawyer living in Eugene, Oregon.

Photo Credits

Cover (left), p. 15 © Biophoto Associates/Photo Researchers, Inc.; cover (right) © LSHTM/Photo Researchers, Inc.; p. 5 © Dr. Gopal Murti/Photo Researchers, Inc.; pp. 7, 9, 12, 19, 20, 23, 26, 27, 35, 37 by Tahara Anderson; p. 13 © Asa Thoresen/ Photo Researchers, Inc.; p. 17 © Royalty-Free/Corbis; p. 21 © 2000–2004 Custom Medical Stock Photo; p. 29 © P. Dayanandan/Photo Researchers, Inc.; p. 30 © Sidney Moulds/Photo Researchers, Inc.; p. 31 © Sheila Terry/Photo Researchers, Inc.; p. 39 © David M. Philips/Photo Researchers, Inc.

Designer: Tahara Anderson; **Editor:** Joann Jovinelly